Lean Six Sigma

The Fundamental and Detailed Approach to
Understand and Master Six Sigma Qualities
and Lean Production Speed for Beginners in
Less than 14 Days Part-1

By Jackson Jenkins

Table of Contents

Introduction to lean six sigma 7

What is Lean Six Sigma? 21

What Is Continuous Improvement? Definition and Tools 43

Embracing Continuous Improvement – Tools and Techniques 48

The Integration of Six Sigma and Manufacturing 70

By reading this document, the reader agrees that under no circumstances is the author responsible for any losses, direct or indirect, which are incurred as a result of the use of information contained within this document, including, but not limited to, — errors, omissions, or inaccuracies.

Introduction to lean six sigma

If you are beginning to find out about the ideas of streamlining a business procedure, you are at a suitable spot. We will acquaint you with the approaches of Lean, Six Sigma, and Lean Six Sigma. Individuals with no involvement with this region can get a thought of what it is about. So, we should start:

Segment I: Lean Methodology

What is Lean?

Lean is a way to deal with a decrease or dispose of exercises that don't increase the value of the procedure. It underlines expelling inefficient strides in a system and making the primary worth included strides. The Lean strategy guarantees high caliber and consumer loyalty.

It helps in

- lessening process duration,

- improving item or administration conveyance time,
- decreasing or wiping out the opportunity of deformity age,
- reducing the stock levels and
- advancing assets for key upgrades, among others.
- It is a ceaseless way to deal with squander evacuation; in this manner, it advances an unbroken chain of enhancements.

What is "Worth"?

We should comprehend what is "Worth" in the above definition on Lean:

Contingent upon the kind of business process and industry setting, the client characterizes "esteem." "Worth" is identified with the client's view of the product(s) or service(s), which the person is happy to pay for.

A procedure is a set of exercises, which change over contributions to yields utilizing assets. In a system, these exercises can be arranged into three sorts. They are:

Non-Value-included action: These exercises don't increase the value of the processor items. They structure inefficient advances. A client doesn't pay for the expenses related to these exercises eagerly. Or maybe, if present exorbitantly, they bring about client disappointment.

Worth included action: These exercises increase the value of the procedure and are fundamental. They improve forms for profitability and quality.

Empowering esteem included movement: These exercises don't increase the value of a client. They are fundamental to the progression of a procedure.

In any procedure, very nearly 80 – 85% exercises are non-esteem, including practices. The point of LEAN methodology is to distinguish them all the while. What's more, utilize explicit lean apparatuses to dispose of or diminish them. Along these lines, Lean improves process effectiveness.

Evacuating Waste:

Lean idea acquires its beginning from TPS – Toyota Production framework. The TPS model ordinarily is appropriate for High Volume Production conditions. In any case, Lean discovers its application in any situation where procedure squanders are seen. Lean can be applied to assembling just as administration enterprises. It causes presumably that Lean, these days, is being received by administration parts with the two arms.

Procedure squanders recognized in Lean Methodology is known as "Muda." Muda is a Japanese group for wastes – presented by the Japanese architect Taiichi Ohno of (Toyota) in the 1960s.

Utilizing the Lean approach, you can evacuate beneath referenced eight sorts of waste ("DOWNTIME" is the abbreviation for the eight squanders).

Squanders

- **Deformities**

The endeavors included investigating for and fixing blunders, messes up through revamps.

- **Overproduction**

Creating more items or administrations that the client needs or downstream procedure can utilize.

- **Pausing**

Inert time made when material, data, individuals, or hardware isn't prepared. It incorporates high employment set up time in assembling. Or then again, too much senior information preparing time in the administration business.

- **In – Utilized Talent**

Not sufficiently utilizing people groups' abilities and inventiveness. Representative strengthening can counter this loss as supported by Japanese quality pioneers.

- **Transportation**

Moving items, hardware, material, data, or individuals are starting with one spot then onto the next, with no worth expansion to certain articles or administration.

- **Stock**

Superfluous/Unwanted legging or capacity of data as well as material (eg, WIP, WIQ – work in the line)

- **Movement**

Unnecessary development of individuals or machines that requires some serious energy and utilizations vitality. It might make weariness workers because of the undesirable event of a body.

- **Additional Processing**

Procedure steps that don't enhance the item or administration, including doing work past a client's detail.

The Five Principles of Lean

These Lean standards can be applied to any procedure to diminish the squanders. They are:

Characterize Value: The client characterizes the estimation of an item or administration. Consequently, the initial step is to distinguish clients. Ask yourself, what does the client esteem? Make sense of the client's desires from your items or administrations. Characterize the procedure exercises into Non-Value included, Value-included, and Enabling worth included.

Guide the worth stream: The worth stream mapping shows the work process ventures for an item or administration. The worth stream mapping distinguishes and wipes out NVA exercises. This, in the end, causes you to decrease the procedure delays and, in this manner, improves the nature of item/administration.

Make Flow: Create a stream to the client by guaranteeing a constant stream framework in delivering items or administration. The stream will advance the procedure to augment process effectiveness.

Set up Pull: build-up to pull approach by meeting framework beat time. The time is the rate at which an item should be prepared to satisfy the client's needs. JIT (Just in time) is an instrument advancing the Pull framework. This guarantees a smooth work process of the procedure with no disturbances. It additionally lessens the stock level.

Look for Continuous Improvement: Finally, you should put predictable endeavors to improve the current business procedures to cook consistently changing client needs. This guarantees disposal of waste and imperfections free items and quality help to clients.

Prologue to some significant Lean devices:

VSM (Value stream mapping): as of now examined, VSM distinguishes process squanders and reasons for these squanders.

Kaizen: It's a consistent improvement approach concentrating on little – little upgrades. It includes the responsibility of down level individuals in the association towards process upgrades, encouraged by subordinates, and upheld by the board.

Without a moment to spare: It's a dismantle way to deal with satisfy client needs as and when it streams from a client.

SMED (Single moment trade of bites the dust): It improves gear changeover time. It chips away at a standard of lessening changeover time to inside ten minutes.

Jab Yoke: It's a mix-up sealing gadget utilized in gathering to caution administrators on deformities or disappointments.

Judoka (Autonomation): Also known as keen computerization. It stops the get-together or creation line if an imperfection happens.

Heijunka: It's the idea of Line Balancing. The point is to appropriate the heap by adjusting generation lines uniformly.

Gemba (Go and See): The point is to go to the real work environment. Watch the procedure and executions continuously with care. Record their perceptions. It's another method to discover process traps.

Kanban: It's a sign framework to oversee stock level. Kanban loads up can be shown and figured out how to see the present stock level consistently. It additionally alarms to the administration to bring consideration over exorbitant stock. Over the top stock ties up the working capital and squares it from gainful use.

Presently we should comprehend about administration approach of Six Sigma.

Segment II: Six Sigma

What is Six Sigma?

Six Sigma is information driven critical thinking system. The attention is on process varieties, and accentuation is given to consumer loyalty. Continues process improvement with low deformities is the objective of this strategy.

The objective of Six Sigma :

The point of Six Sigma is to make a procedure powerful with - 99.99996 % imperfection free. This implies a six-sigma process creates in 3.4 defects per million chances or less subsequently.

Six Sigma is an organized, critical thinking approach. Critical thinking in Six Sigma is finished utilizing the DMAIC system. There are five phases in this structure. They are

- Characterize,
- Measure,
- Break down,
- Improve,

- Control.

- Portrayal of Phase

- Characterize

In this stage, venture goals are sketched out. A venture contract is a significant segment of this stage. A task contract is a diagram record for a six-sigma venture. An ordinary agreement contains the accompanying data:

- Business case

- Issue explanation

- Objective explanation

- Venture scope

- Assets

- Timetables

Assessed benefits

This contract gives an outline of a six-sigma venture and is endorsed by top administration to give a thumbs up to six sigma venture.

Measure

Procedure factors are estimated at this stage. Procedure information is gathered. The pattern is acquired, and measurements are contrasted and last execution measurements. Procedure ability is acquired.

Break down

Underlying driver investigation is done at this stage. Complex investigation instruments are used to distinguish the underlying drivers of an imperfection. Devices like histograms, Pareto graphs, fishbone charts are utilized to identify the underlying drivers. Theories tests are directed to check and approve main drivers, Viz Regression test, ANOVA test, Chi-square, and so forth.

Improve

When the last main drivers are recognized, arrangements should be framed to improve the procedure. Steps to identify, test, and execute the answers for wipe out main drivers are a piece of this stage. Recreation examines, Design of trials, Prototyping is a portion of the strategies utilized here to improve and amplify process execution.

Control

After actualizing the arrangements, the presentation of the methods must be recorded. A control framework must be set up to screen the presentation post improvement. What's more, a reaction plan is created to deal with arrangement disappointment. Procedure institutionalization through Control plans, and work directions is commonly a piece of this stage. Control outlines show the procedure execution. Venture benefits are talked about and checked against assessed one. The principal reason for this stage is to guarantee to hold the increases.

Area III: Lean Six Sigma

What is Lean Six Sigma?

Characterizing Lean Six Sigma:

ASQ (The American Society for Quality) states,

"Lean Six Sigma is a reality-based, information-driven way of thinking of progress that qualities imperfection counteractive action over deformity location. It drives consumer loyalty and primary concern results by lessening variety, waste, and process duration, while advancing the utilization of work institutionalization and stream, in this way making an upper hand. It applies anyplace variety and waste exist, and each worker ought to be included."

Lean Six Sigma joins the techniques of Lean and Six Sigma. Lean standards help to lessen or kill process squanders. Six Sigma centers around variety - decrease in the process. In this way, the rules of Lean Six Sigma helps to improve the proficiency and nature of the procedure.

Why is Lean Six Sigma picking up the significance in the present situation?

The present condition is dynamic. Lean or six sigma approach in this unique condition can't carry the maximum capacity to enhancements whenever applied in seclusion. Joining of Lean and

Six Sigma guarantees remarkable improvements. In this administration approach, generally, the lean strategy is utilized first to evacuate the loss in a procedure. Afterward, the Six Sigma apparatuses are being used to improve process varieties. Be that as it may, these two techniques go connected at the hip in the present time. A definitive goal is to enhance forms by lessening variety and taking out the waste. It's a consistent improvement process, where Lean techniques and Six Sigma draws near, both go ahead during PDCA. The degree of approaches may vary contingent on process complexities or improvement looked for. The blend of these two techniques creates streamlined procedures with top-notch and results. It improves the main concern benefits and helps meeting business objectives.

The coordinated Lean Six Sigma the executive's approach is being utilized crosswise over divisions and ventures. It elevates to notable changes in the association's presentation. Lean Six Sigma prompts getting a charge out of upper hands in different organizations on the planet. They can be an item or administration arranged organizations. The LSS strategy improves procedures and makes them productive. The way to progress is the executive's support, worker commitment, and promise to enhancing consumer loyalty.

More or less, Lean procedure goes for squander decrease in process, while six sigma goes for a reduction of procedure variety. Notwithstanding, both the methodologies go inseparably to understand the maximum capacity of procedure upgrades. An incorporated method of lean six sigma helps improving procedure effectiveness, upgrading assets, and expanding consumer loyalty while improving benefits and shortening cost.

Expectation, presently, you comprehend the contrasts between these three administrations draws near. They have their advantages when applied to various business forms. They improve the nature of existing procedures and make you a superior director.

Lean Six Sigma is a strategy that depends on a community-oriented collaboration to improve execution by efficiently evacuating waste and lessening variety. It consolidates lean assembling/lean endeavor and Six Sigma to dispose of the eight sorts of waste (Muda): Defects, Over-Production, Waiting, Non-Utilized Talent, Transportation, Inventory, Motion, and Extra-Processing.

Lean Six Sigma diminishes process deformities and waste, yet also, it gives a system to by and sizeable hierarchical culture change. By presenting Lean Six Sigma, the mentality of representatives and administrators change to one that spotlights on development and persistent improvement through procedure advancement. This adjustment in culture and the mindset of an association expand productivity and build gainfulness.

To effectively execute Lean Six Sigma, a mix of devices from both lean assembling and Six Sigma must be utilized. A portion of these devices incorporates kaizen, esteem stream mapping, line adjusting, and visual administration. Squander

Fujio Cho of Toyota characterizes squander as "something besides the base measure of hardware, materials, parts, space, and laborers time, which are significant to increase the value of the item."

Various kinds of waste have been characterized:

Imperfections: A deformity is an item that is pronounced unfit for use. This requires the subject to either be rejected or revamped, costing the organization time and cash. Models incorporate a thing that is scratched during the creation procedure and off base get together of an issue because of hazy guidance.

Over-Production: Over-generation alludes to the item that is made in abundance or made before it is required. Things ought to be created as they are needed after the Just-in-time producing theory in Lean. Models incorporate making pointless reports and overproduction of an item before a client has mentioned it.

Pausing: Waiting includes delays in process steps and is part of two unique classifications: hanging tight for material and hardware and inert gear. Models incorporate sitting tight for approval from an unrivaled, hanging tight for an email reaction, hanging tight for material conveyance, and moderate or defective equipment.

Non-Utilized Talent: Non-Utilized Talent alludes to the misuse of human potential and ability and is the freshest expansion to the eight squanders. The fundamental driver of this waste is when the board is isolated from workers. At the point when this happens, workers are not allowed the chance to give criticism and proposals to directors to improve the procedure stream and creation endures. Models incorporate inadequately prepared representatives, absence of motivators for workers, and setting representatives in occupations or places that don't use the entirety of their insight or expertise.

Transportation: Transportation is pointless or over the top development of materials, items, individuals, gear, and devices. Traffic increases the value of the issue and can even prompt item harm and imperfections. Models incorporate moving pieces between various useful territories and sending the overloaded stock back to an outlet distribution center.

Stock: Inventory alludes to an overabundance in items and materials that aren't yet prepared. This is an issue because the article may get outdated before the client requires it, putting away the stock costs the organization time and cash, and the plausibility of harm and imperfections increments after some time. Models incorporate overabundance completed products, completed merchandise that can't be sold, and broken machines dispersed on the assembling floor.

Movement: Motion is pointless development by individuals. The unnecessary movement sits around idly and expands the opportunity of damage. Models incorporate strolling to get devices, going after materials, and strolling to various pieces of the assembling floor to finish multiple assignments.

Extra-Processing: Extra-preparing is accomplishing more work than is required or essential to finish an assignment. Models incorporate twofold entering information, superfluous strides underway, pointless item customization, and utilizing higher

accuracy hardware than would generally be appropriate.

History

The historical backdrop of lean includes illuminators, for example, Toyoda, Ford, Ohno, Taylor, and numerous others. Find out about their accomplishments and commitments.

At the point when we talk about lean, the first name that strikes our brain is Toyota. In any case, it is significant that the historical backdrop of thin began path back in the 1450s in Venice. From that point, the primary individual who incorporated the idea of lean in the assembling framework was Henry Ford. Further, in 1799, Eli Whitney accompanied the concept of tradable parts. At that point in 1913, Henry Ford propounded the progression of generation by exploring different avenues regarding the exchanging and development of various components to accomplish institutionalization of work. In any case, there was a restriction to

Ford's framework that it needed assortment and was applied to just a single particular.

It was merely after World War II in 1930s Toyota propelled from

Ford's progression of generation idea and concocted the Toyota Production System. The reason for this new framework is to change the concentration from the utilization and usage of individual machines to the workstream from the complete procedure. The Toyota Product framework targets diminishing the expense of creation, improving the nature of items, and expanding the throughput times with the goal that the dynamic client needs are met. A portion of the means that the framework consolidated are the sight-measuring of the machines keeping in thought the required volume of generation, automatic highlights of device with the goal that the nature of the fabricated items is upgraded, sequencing the tools according to the procedure, grow fast advances, so the assembling of numerous parts in relative little volume gets conceivable and keeping a substantial correspondence of the necessities of the regions between the means of the procedure.

Critical Voices in the History of Lean

The idea of lean administration was resuscitated continuously with the changing occasions and needs of the business. With this dynamism in the new condition, various defenders of lean assembling made considerable commitments in the field of lean administration.

Frederick Taylor

From the get-go during the 1890s, the dad of logical administration, Frederick Taylor, examined intently in the work techniques and laborers at the production line level. After his supervision, he propounded ideas like institutionalization of work, time studies, and movement examine, to accomplish effectiveness in the work strategies, procedures, and activities. Be that as it may, he disregarded the conducting part of the work, which welcomed numerous reactions against him.

The Principles of Scientific Management: Outlines the establishment for present-day association and choice hypothesis. He does this by portraying the difficulty: laborers harbor fears that higher individual efficiency will, in the end, lead to fewer employments. Taylor's recommendation is to disparage this dread by giving motivations to laborers and re-encircling the customer, investor, and specialist relationship. An excellent and enlightening read for anybody keen on effective administration rehearses.

Henry Ford

Beginning in 1910, Henry Ford spearheaded the well-known assembling system, wherein every one of the assets utilized at the assembling site-individuals, machines, hardware, apparatuses, and items were orchestrated in such a way, that a consistent progression of creativity is encouraged. Most punctual American promoters of waste decrease (LEAN). He achieved massive accomplishment by this procedure in assembling the Model T vehicle and even turned into the most extravagant man on the planet.

In any case, as the world started to change, Ford couldn't change the work strategies and bombed when the interest of the market was to include new models, hues, and assortments to the items. At last, during the 1920s, the worker's guilds and the thing multiplication consumed the accomplishment of Ford, and it was by mid-1930s that General Motors commanded the car advertise.

Sakichi Toyoda

Sakichi Toyoda built up the Toyoda turning and weaving organization in the year 1918. Sakichi Toyoda was one of the underlying donors of progress towards the celebrated Toyota Production framework that targets wiping out every one of the squanders, by propounding the Jidoka idea. Jidoka-'mechanization with a human touch' signifies to encourage quality at source. He imagined the programmed loom in 1896 that subbed the manual work as well as introduced the ability to make decisions into the machine itself. The framework improved the work adequacy and effectiveness by moderating the item's deserts and related inefficient work rehearses. The guideline was prompted an early discovery of a variation from the no simple halting of the machine or procedure on the recognition of the issue, the quick obsession of the anomaly, and even examined the main driver of the problem.

Kiichiro Toyota

Kiichiro Toyota was the originator and second leader of Toyota Motor Corporation. He was the child of Sakichi Toyoda, and later in 1937, he established the Toyota Motor Corporation. He took forward his dad's idea of Jidoka and built up his way of thinking about without a moment to spare (JIT) idea in assembling. He paid to visit Ford's plant in Michigan to comprehend the progression of the mechanical production system idea and afterward proposed the Toyota Production framework. The new structure went for right, estimating the machines regarding the real volume required and presented botch sealing with the goal that quality is guaranteed. Appropriate sequencing of work forms is finished.

Taichi Ohno

Perhaps the greatest accomplishment of Taichi Ohno was to coordinate the Just-in-Time framework with the Jidoka System. After his visit to America to ponder Ford's strategies in 1953, he got exceptionally motivated. He comprehended the future needs of the buyers that they will choose the required items from the racks and how the things renewed. This has motivated him to fabricate the useful Kanban framework. He even rehearsed the Dr. Edwards Deming strategy to consolidate quality at each progression of the procedure from structure to aftersales administrations to the purchasers. This was repeated and brought at the floor level by Ohno, who incorporated this way of thinking with the Kiirocho's in the nick of time idea and standard of Kaizen. He is considered as the genuine modeler of the 'Toyota Production System.'

The Toyota Production framework depends on various ideas, which are pull framework, disposal of waste, Quick Die changes (SMED), non-esteem included work, U-molded cells, and one-piece stream. The force framework characterizes the flow of the material in the middle of various procedures as controlled by the requirements of the clients. The organization rehearses the force framework by utilizing a Kanban framework, which gives a sign to the client that the apparatuses are accessible to be moved to the following procedure in the grouping.

The Toyota Production System likewise distinguishes the waste, named as Muda, and perceives that waste is whatever doesn't increase the value of the clients. Squanders are of seven kinds; that is over-generation, stock waste, abandons, pausing, movement, over-preparing, and transportation and dealing with. The framework targets recognizing and taking out these squander to encourage productivity and adequacy in the creation framework.

Another strategy embraced by the organization is the Quick pass on changes (Single Minute Exchange of Dies), which targets improving the stream (Mura) of creation. The idea depends on the reason that the instruments and changeovers should take short of what one moment (single digit) at the most extreme. During the 1950s and 1960s, the organization experienced the nearness of bottlenecks at the vehicle body forming presses. The main driver was distinguished to be the high changeover times that expanded the part size of the creation procedure and drives up the generation cost. Toyota actualized the SMED by putting the accuracy estimation gadgets for the exchange of substantial weight bites the dust on enormous exchange stepping machines used to create the body of the vehicles.

Further, the long creation lines at Toyota were wrapped utilizing U-molded cell design that encourages lean assembling. This expands the effectiveness of the laborers to work at numerous machines one after another. The Toyota Production System (TPS) rehearses the one-piece stream, whereby the past activity controls the need for the sequential procedure. Each piece is created, in turn, restricted to large-scale manufacturing. Toyota puts a solitary piece between various workstations with the benefit of least change in process duration and least holding uptime. This would encourage ideal harmony between multiple activities and alleviate over-generation, which is viewed as one of the significant squanders.

The Toyota Way: In production lines far and wide, Toyota reliably makes the most excellent autos with the least deformities of any contending maker while utilizing fewer worker hours, less close by stock, and a large portion of the floor space of its rivals. The Toyota is the first book for a general group of spectators that clarifies the administration standards and business reasoning behind Toyota's overall notoriety for quality and unwavering quality.

Shigeo Shingo

Dr. Shigeo was an industrialist engineer, and a significant specialist at Toyota, that effectively helped the organization accomplish lean assembling. He aced the Kaizen idea. He comprehended the achievement of lean assembling by incorporating individuals with compelling and effective procedures. In 1960, he built up the SMED framework with the intent to accomplish zero quality imperfections.

A Revolution in Manufacturing: The SMED System. Composed by the mechanical specialist who created SMED (single-minute trade of kick the bucket) for Toyota, A Revolution in Manufacturing gives a full review of this incredible without a moment to spare generation apparatus. It offers the most complete and point by point directions accessible anyplace for changing an assembling domain in manners that will accelerate the generation and make little parcel inventories attainable. The creator dives into both the hypothesis and practice of the SMED framework, clarifying basics just as strategies for applying SMED. The widely praised content is bolstered with several representations and photos, only as twelve section length contextual analyses.

Zero Quality Control: the Poke-Yoke framework. A mix of source investigation and misstep sealing gadgets is the primary technique to get you to zero imperfections. Shigeo Shingo gives you how this demonstrated framework for decreasing mistakes turns out the essential items in the most-brief timeframe. Shingo gives 112 specific instances of poka-burden advancement gadgets on the shop floor, the vast majority of them costing under $100 to execute. He likewise talks about assessment frameworks, quality control circles, and the capacity of the executives concerning the review.

A Study of the Toyota Production System: Here is Dr. Shingo's great mechanical building justification for the need for the procedure based on operational upgrades in assembling. He clarifies the essential components of the Toyota generation framework, analyzes creation as an efficient system of processes and tasks, and afterward talks about the instrument important to make JIT conceivable in any assembling plant.

Kaizen and the Art of Creative Thinking: Dr. Shingo uncovers how he showed Toyota and other Japanese organizations, the specialty of recognizing and tackling issues. Numerous organizations in the West are attempting to copy Lean; however, not many can do it. Why not? Conceivably, because we in the West don't perceive, create, and bolster the imaginative capability of each specialist in tackling issues. Toyota makes all representatives issue solvers. Dr. Shingo gives you the instruments to do it.

1980s-2000s

What has today to become Lean Motorola can follow six Sigma in the United States in 1986. Six Sigma was created inside Motorola to rival the Kaizen (or lean assembling) plan of action in Japan. Because of Six Sigma, Motorola got the Malcolm Baldridge National Quality Award in the year 1988.

During the 1990s, Allied Signal enlisted Larry Bossidy and presented Six Sigma in overwhelming assembling. A couple of years after the fact, General Electric's Jack Welch counseled Bossidy and started Six Sigma at General Electric. Now, Six Sigma turned out to be all the more generally acknowledged and known in the assembling scene.

During the 2000s, Lean Six Sigma forked from Six Sigma and turned into its very own special procedure. While Lean Six Sigma created as a particular procedure of Lean Six Sigma, it likewise fuses thoughts from lean assembling, which was created as a piece of the Toyota Production System during the 1950s.

2000s-2010s

The original idea of Lean Six Sigma was made in 2001 by the book titled Leaning into Six Sigma: The Path to Integration of Lean Enterprise and Six Sigma by Barbara Wheat, Chuck Mills, Mike Carnell. The book was created as a guide for supervisors of assembling plants on the most proficient method to join lean assembling and Six Sigma so as to improve quality and process duration in the plant drastically. Wheat, Mills, and Carnell portray the tale of an organization that was wary about actualizing Lean Six Sigma, yet because of doing so had the option to effectively improve the quality and effectiveness in all parts of the business.

In the mid-2000s, Six Sigma standards ventured into different parts of the economy, for example, Healthcare, Finance, Supply Chain, and so on. While various divisions of the economy sell diverse "items" and have distinctive "clients," Lean Six Sigma standards can, at present, be applied with slight changes in wording and procedures.

What Is Continuous Improvement? Definition and Tools

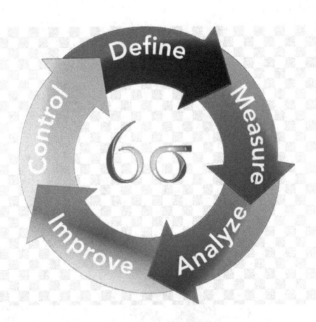

In Lean, continuous improvement resembles a religion. Although it appears to be a straightforward thing to accomplish, pioneers, and groups, who are curious about procedure improvement strategies are making some hard memories continuing it.

To execute this outlook, you have to have a reasonable comprehension of what precisely is a consistent improvement, what standards you have to pursue and check the entire prescribed procedures.

The Continuous Improvement Model

The term consistent improvement can be exceptionally dynamic if not put in a particular setting. Clarified in the blink of an eye, it is a ceaseless take a stab at flawlessness in all that you do. In Lean administration, consistent improvement is otherwise called Kaizen.

Kaizen began in Japan not long after the finish of the Second World War. It increased monstrous prominence in assembling and got one of the establishments of Toyota's ascent from a little vehicle producer to the biggest carmaker on earth.

With regards to the Lean procedure, persistent improvement looks to improve each process in your organization by

concentrating on upgrading the exercises that create the most incentive for your client while expelling whatever number waste exercises as could reasonably be expected.

There are three kinds of waste in Lean:

- Muda – The seven squanders
- Mura – The misuse of lopsidedness
- Muri – The abuse of overburden

Muda comprises of 7 significant procedure squanders transport, stock, movement, pausing, overproduction, over-handling, surrenders.

Evacuating every one of them is incomprehensible, yet concentrating on limiting their negative consequences for your work is pivotal for the fruitful usage of persistent improvement.

Mura is brought about by lopsidedness or irregularity in your procedure. It is liable for a large number of the seven squanders of Muda. Mura prevents your undertakings from streaming easily over your work procedure and in this way, hinders you from arriving at the ceaseless stream.

Muri is a significant issue for organizations that apply push frameworks. At the point when you dole out an excess of work to your group, you place excessive weight on both your group and procedure.

Muri is generally an aftereffect of Mura, and on the off chance that you need constant improvement to turn out to be a piece of your way of life, you have to concentrate on disposing of those squanders.

Embracing Continuous Improvement – Tools and Techniques

Understanding the hypothesis behind it is the initial phase in applying a consistent improvement to your administration culture. To set yourself up for constant improvement, you have to make a reasonable situation inside your organization.

In Lean administration, there are three significant methodologies for accomplishing consistent improvement:

Plan-Do-Check-Act (PDCA)

The model Plan-Do-Check-Act is the most prevalent methodology for achieving continual development.

Otherwise called the Deming circle (named after its author, the American specialist William Edwards Deming), it is an endless cycle that intends to assist you with improving additionally dependent on accomplished outcomes.

It was first created for quality control, yet in time turned into an instrument for accomplishing continual improvement.

In the arranging stage, you have to set up destinations and procedures essential to convey brings about understanding with the average yield (the objective or objectives).

Setting yield desires is a vital aspect for accomplishing persistent improvement because the exactness of the objectives and their culmination is a significant piece of the way toward improving.

It is prescribed to begin a little scale so you can test the impacts of the methodology.

The subsequent stage is "Do." It is direct as you have to execute what you've set down during the arranging venture of the procedure.

After you've finished your targets, you have to check what you've accomplished and contrast it with what you've anticipated. Assemble, much information as could be expected and think about what would you be able to improve in your procedure so you can accomplish more prominent outcomes next time.

On the off chance that the investigation shows that you've improved contrasted with your past venture, the standard is refreshed, and you have to go for a stunningly better presentation next time.

If you've neglected to improve or have even accomplished more terrible outcomes contrasted with the past, the standard remains as it was before you began your last undertaking.

Underlying driver Analysis

Underlying driver Analysis (RCA) is a method rehearsed in Lean administration that enables you to accomplish Kaizen by indicating you the main driver of the issues in your procedure.

It is an iterative practice that drills down into an issue by investigating what caused it until you arrive at the base of what is creating a negative impact. It very well may be viewed as root just if the last negative impact is forestalled for good after the reason is evacuated.

To apply RCA for continuous improvement, you have to play out an exhaustive investigation of the issue.

For instance, suppose that you are driving a product improvement group. At the point when you discharged the most recent update of your item, your help group was barraged with bug reports from clients.

You start to search for the underlying driver beginning from the highest point of the issue.

You examine how your QA group considered this to occur and find that they neglected to run all the necessary tests on the product.

A short time later, you investigate what caused that and discover that the improvement group gave them the highlights that should have been discharged ultimately.

Investigating the reason for that, you discover that the designers completed most of the highlights directly before they submitted them for quality confirmation.

Delving into the reason for that, you discover that your advancement group took additional time than you have wanted to build up the highlights in the lead position.

Researching the explanation for that, you find that your group was wasteful because every engineer took a shot at a couple of highlights at the same time, and hence, as opposed to giving highlights individually to QA they presented a cluster that was too huge to even think about processing without prior warning.

Examining why this occurred, you understand that you haven't set any guidelines on the measure of work that can be in progress at the same time and didn't guarantee the uniformity of your procedure.

Arriving at this point, you get to the end that the main driver of the bug issue is Mura (the misuse of lopsidedness).

To accomplish constant improvement, we propose you break down the primary driver of every issue and examination with arrangements.

Frequently, issues may end up being more mind-boggling than you might suspect, and the RCA would require a couple of emphases before keeping the negative impact from consistently happening once more.

If you don't know how to play out the main driver examination, we recommend you investigate the 5 Whys for deciding underlying drivers.

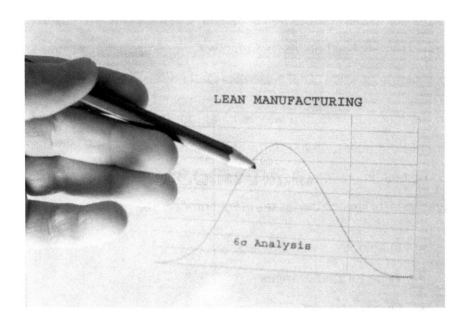

Applying Lean Kanban

To ceaselessly improve your procedure, you have to have a reasonable representation of what should be improved.

On the off chance that you need permeability, you'll have the option to improve now and again yet won't have the opportunity to spot side effects of an issue before it is past the point of no return.

At the point when Toyota was searching for an approach to do that, they created Kanban as a framework for improving the work process proficiency of the generation procedure.

In the long run, Kanban was adjusted for information work and figured out how to help a large number of groups to accomplish consistent improvement. The strategy depends on six center practices for limiting the losses in your procedure:

- Imagine your work process
- Wipeout interferences
- Oversee stream
- Make process approaches unequivocal
- Make criticism circles
- Improve cooperatively

To imagine your work process, the strategy depends on whiteboards for mapping each progression of your procedure. The board is partitioned by vertical lines framing segments for the various stages.

A first Kanban board comprises three segments: Requested, in progress, Done.

Each undertaking that your group is taking a shot at is facilitated on a Kanban card (initially as a post-it note) and needs to go through every one of the phases of your work process to be viewed as complete.

Kanban sheets enable you to screen the equality of your procedure and can be a genuine weapon for limiting Mura.

Moreover, they show you the measure of work that each individual in your group has and can assist you with forestalling overburden (Muri) by enabling you to assign undertakings as needs are to the limit of your group.

To wrap things up, you can screen the pace at which work is advancing over your work process and accomplish continual improvement of your work process effectiveness.

For disposing of interferences, Kanban depends on restricting the work that can be in progress all the while. The objective is to dispense with performing multiple tasks, which is just a steady setting switch among assignments and only damages profitability.

With the assistance of Kanban, you can deal with the progression of work in your procedure. To guarantee a smooth process, you should know about where work stalls out and make a move to ease the bottlenecks in your system. Along these lines, you can explore different avenues regarding the many strides of your work process and continue improving continually.

In Lean administration, constant improvement is a gathering movement. In this manner, you have to ensure that your group comprehends the shared objective and why their piece of the procedure is significant.

By making process approaches straightforward, you'll urge your colleagues to assume greater liability and take responsibility for the process.

For positive change to occur, there should be a steady progression of information among you and your group.

The Kanban load up itself is an extraordinary criticism circle generator since it makes it visible who is doing what whenever.

In blend with the broadly embraced act of holding day by day stand up gatherings between the group, you can continuously improve data sharing between people.

Different procedures that are a piece of the constant improvement arms stockpile are the Gemba walk and the A3 report. The A3 news is an organized methodology that causes you to manage critical thinking issues, while the Gemba walk urges you to take a quick trip and see where the genuine work occurs. Both are incredibly valuable, and they can assist you in finding tricky parts in your work process.

Primary concern

Kaizen is an endless mission for flawlessness; however, you'll begin feeling the advantages of persistent enhancement for your business when your entire group takes it by heart.

Kanban and the different continuous improvement apparatuses can help your bounty with that because your group will get heaps of information about procedure improvement and work process the executives. Subsequently, every individual will increase a superior comprehension of how your procedure functions and how it very well may be improved.

Depiction

Lean Six Sigma is a synergized administrative idea of Lean and Six Sigma. Lean generally centers around the end of the eight sorts of waste/Muda named abandons, over-generation, pausing, non-used ability, transportation, stock, movement, and extra-handling. Six Sigma tries to improve the nature of procedure yields by distinguishing and expelling the reasons for abandons (mistakes) and limiting inconstancy in (assembling and business) forms. Together, Lean plans to accomplish nonstop stream by fixing the linkages between process steps while Six Sigma centers around decreasing procedure variety (in the entirety of its structures) for the procedure steps along these lines, empowering a fixing of those linkages. To put it plainly, Lean uncovered wellsprings of procedure variety, and Six Sigma intends to lessen that variety allowing a righteous cycle of iterative enhancements towards the objective of the persistent stream.

Lean Six Sigma utilizes the DMAIC stages like that of Six Sigma. The five steps incorporate Define, Measure, Analyze, Improve, and Control. The five stages used in Lean Six Sigma are intended to distinguish the primary driver of wasteful aspects and works with any procedure, item, or administration that has a lot of information or quantifiable qualities available. The DMAIC toolbox of Lean Six Sigma includes all the Lean and Six Sigma instruments.

IASSC is an association that gives Lean Six Sigma related to preparing, tutoring, instructing, and counseling services. With IASSC, intrigued experts can demonstrate their insight into Lean Six Sigma with various tests and affirmations. The multiple degrees of statements are isolated into belt hues, like judo. The most elevated level of confirmation, implying profound information on Lean Six Sigma standards, is a dark belt. Underneath the colored belts are the green and yellow belts. For every one of these belts, levels range of abilities is accessible that portray which of the general Lean Six Sigma devices are required to be part at a specific Belt level. These ranges of abilities give a nitty-gritty depiction of the learning components that a member will have procured after finishing a preparation program. The fields of skills reflect elements from Six Sigma, Lean, and different procedure improvement techniques like the hypothesis of limitations (TOC) all out gainful support (TPM). To accomplish any of the accreditation levels, an administered test must be passed that remembers different inquiries for Lean Six Sigma and its applications.

Lean Six Sigma is a group centered administrative methodology that tries to improve execution by disposing of waste and deformities. It consolidates Six Sigma techniques and devices and the lean assembling/lean endeavor theory, endeavoring to wipe out misuse of physical assets, time, exertion, and ability while guaranteeing quality underway and hierarchical procedures. Basically, under the precepts of Lean Six Sigma, and utilization of assets that don't make an incentive for the end client is viewed as a waste and ought to be dispensed with.

Separating Lean Six Sigma

Lean Six Sigma can see its sources in the United States during the 1980s as a mix of the board standards and procedures that began in Japan. With an end goal to contend with Japan's better items, U.S. directors embraced some Japanese assembling rules that concentrated on diminishing waste as non-esteem, including activities. During the 1990s, such standards were received by huge U.S. producers. Michael George and Robert Lawrence Jr. presented Lean Six Sigma in their 2002 book Lean Six Sigma: Combining Six

Sigma with Lean Speed as a mix and refinement of the lean venture and Six Sigma fundamentals.

Lean Six Sigma Tenets

The "lean" idea of the executives loans its emphasis on the decrease and end of eight sorts of waste known as "Personal time," which is a shortening of deformities, overproduction, pausing, non-used ability, transportation, stock, movement, and extra-preparing. "Lean" alludes to any strategy, measure, or device that aids in the ID and end of waste.

The term Six Sigma alludes to devices and systems that are utilized to improve producing forms. It was presented by a designer at Motorola in 1986 and animated by Japan's Kaizen model. The organization trademarked it in 1993. It plans to improve forms by recognizing and wiping out the reasons for imperfections and varieties in business and assembling structures. Six Sigma's DMAIC stages are used in Lean Six Sigma. The abbreviation represents characterize, measure, investigate, improve and control and alludes to an information-driven strategy for improving, upgrading, and balancing out business and assembling forms.

How Lean and Six Sigma meet up

Lean Six Sigma uses ideas from both Lean and Six Sigma to cut generation costs, improve quality, accelerate, remain aggressive, and set aside cash. From Six Sigma, organizations profit by the diminished, minor departure from parts. Additionally, Lean sets aside money for the organization by concentrating on the sorts of waste and how to lessen squander. The two procedures meet up into Lean Six Sigma, making a very much adjusted and composed answer for set aside cash and produce better items.

Albeit Lean and Six Sigma are various procedures, they are corresponding and share numerous likenesses that enable them to stream together consistently. To begin with, both Lean and Six Sigma stress the way that the client characterizes the estimation of an item or administration. This implies when procedures are analyzed, the significance or need of steps in the process ought to be inspected through the eyes of the client. Additionally, Lean and Six Sigma use process stream maps to all the more likely comprehend the progression of creation and distinguish any squanders.

Moreover, both depend on information to figure out which zones of production need improvement in proficiency and to gauge the accomplishment of upgrades. At last, because of executing Lean and Six Sigma, effectiveness commonly improves, and variety diminishes. Skill and range go connected at the hip, with progress in one bringing about an improvement in the other.

Lean and Six Sigma have numerous similitudes; however, are various procedures and were created for multiple purposes. The primary distinction between the two strategies is issued recognizable proof. While Lean centers the issue of wastefulness around the eight squanders, Six Sigma centers around distinguishing wellsprings of variety to diminish wastefulness. Furthermore, Lean and Six Sigma utilize various apparatuses. While Lean uses more information perception instruments, Six Sigma utilizes progressively numerical and expository centered devices.

The similitudes among Lean and Six Sigma take into consideration concurrent usefulness on a similar item or procedure, while their disparities permit the advantage of having a considerable measure of logical apparatuses at the one's transfer.

Lean and six-sigma are industry-perceived ways of thinking for business improvement. At the point when applied to the production network, they can drive effective outcomes. While six-sigma is profoundly process-centered, the production network is centered around the whole worth stream, from procedure to process. This can display difficulties in applying six-sigma in an inventory network condition. Lean six sigma consolidates both lean and six sigma strategies together to wipe out waste and enhance forms. This should be possible through diminishing waste brought about by transportation, stock, movement, pausing, overproduction, and over-handling, just as limiting deformities in fabricated items.

How do the two procedures vary?

Lean and six-sigma has commonly been depicted as two distinct things – thin being centered around waste and six-sigma being centered around quality. Lean considers waste to be whatever doesn't include esteem, while six sigma sees squander as a procedure variety. Lean doesn't quantitate the nature of a procedure as six-sigma does by utilizing the image sigma/σ to assign the array of the procedure yield.

How lean and six-sigma cooperate?

In principle, lean and six-sigma appear to be irrational, yet they do cooperate. This is because the two of them can make positive monetary upgrades to an association through a degree of profitability.

The 'sweet spot' for lean and six-sigma is to utilize them together as one technique to accomplish operational greatness. With attention on procedure, individuals, and reason, lean six-sigma can assist us with achieving a worth stream that conveys the most elevated an incentive to the client and the least conceivable expense and highest caliber. It gives the upper hand and separation that such vast numbers of organizations require.

Lean Six SigmaBelt for Supply Chain - Blended Course

Results-situated, the new Lean Six Sigma Green Belt mixed course shows experts how to apply lean six sigma to the store network to expand client esteem and at last, create estimated business results. Through self-guided online coursework, virtual training sessions, and an improvement venture, understudies will gain proficiency with the abilities, information, and instruments reliable with industry desires for a guaranteed Lean Six Sigma Green Belt.

The Integration of Six Sigma and Manufacturing

The Lean Manufacturing and Six Sigma techniques are progressively being executed together and what we have today is the unified work of both, and organizations have come to comprehend that their coordination makes it conceivable to exploit the qualities of the two procedures, turning into an extensive and successful, reasonable for taking care of different kinds of issues identified with the improvement of systems and items. Routine administration, process institutionalization, and the investigation of times and developments to kill squander are critical highlights of Lean Manufacturing, while at the same time finding the underlying driver for critical thinking requires further extending and examination in Six Sigma. Lean and Six Sigma can be seen as helpful apparatuses for the activity of the frameworks of progress, development, and routine administration that coordinate the arrangement of business the executives. The organizations have actualized Lean Manufacturing with the point of improving the end of waste in the procedures. Organizations utilizing Six Sigma have discovered that by choosing undertakings and doling out them to groups, after observing, the outcomes would show up. Organizations that execute Lean Six Sigma frequently consciousness of the groups, looking for ventures from changed degrees with the focal point of improving the structure of procedures and accomplish the outcomes.

Why coordinate?

The Lean Manufacturing philosophies are progressively being executed together and what we have today is the unified work of both, and organizations have come to comprehend that their mix makes it conceivable to exploit the qualities of the two methodologies, turning into a far-reaching and compelling, reason for taking care of different kinds of issues identified with the improvement of procedures and items.

Routine administration, process institutionalization, and the investigation of times and developments to dispense with squander are critical highlights of Lean Manufacturing, while at the same time finding the main driver for critical thinking requires further extending and examination in Six Sigma.

Lean and Six Sigma can be seen as helpful apparatuses for the activity of the frameworks of progress, development, and routine administration that incorporate the arrangement of business the board. Thus, they are viewed as reference models. Santos et al. call attention to that dependent on a proof about the capability of best practice draw near, various entertainers worldwide have been proposing and advancing diverse reference models. When all is said in done, these models fill in as references for leaders in setting up practices to be utilized in activities and hierarchical procedures for the most part connected with grants, endorsements, or consultancies.

The organizations have executed Lean Manufacturing with the point of improving the end of waste in the procedures. Organizations utilizing Six Sigma have discovered that by choosing ventures and relegating them to groups, after an observing, the outcomes would show up. Organizations that actualize Lean Six Sigma regularly familiarity with the groups, looking for ventures from changed extensions with the focal point of improving the structure of procedures and accomplish the outcomes.

After some time, organizations have understood that these techniques are corresponding and from that point forward, much has been expounded on consolidating Lean with Six Sigma as a procedure improvement approach, utilizing best practices from each. For sure, creators like George, the association of these two philosophies, augments the estimation of the organization.

There are numerous approaches to join these two procedures. As indicated by Corrêa and Gianesi, in the Western world, there has been a developing development to perceive the critical job of assembling in improving the generation procedure and lessening its expenses. Bendel says that the route forward for actualizing Lean Six Sigma depends mostly on the issues the organization is presently confronting and the idea of its business, just as the goals of the organization and its representatives.

Lean assembling

The idea of Lean Production rose in Japan after World War II made by Japanese Toyota's Eiji Toyoda and Taiichi Ohno, intending to conquer the test of reducing expenses while delivering little amounts of numerous kinds of autos.

The meaning of the Lean Production framework is given by Ohno (1988) as:

"The end of waste and pointless components to lessen costs; the fundamental thought is to deliver just what is essential at the vital minute and in the amount required."

The general perspective on the referred to the creator is that he thinks about seven kinds of generation misfortunes:

- overproduction,
- pause,

- transport,
- super handling,
- dealing with,
- faulty items, and
- stock.

Even though it started in assembling, the Lean idea can likewise be sent in a few authoritative zones.

By adjusting exercises that make an incentive by taking out waste, the worth stream is progressed easily and rapidly, as indicated by the client's solicitation and not as per the maker.

Lean Manufacturing looks for process improvement by streamlining its stream, taking out waste, and underlining gains in speed and productivity.

Womack et al. present the five essential rules that can be utilized as a system for an association to execute the Lean approach, these being:

Worth: Precisely determining the value is the beginning stage for Lean reasoning. The quality is characterized distinctly by the end client. Nonetheless, the association must recognize what creates this incentive for the client. Deciding the worth and describing the item, the following stage is to determine the objective cost dependent on the assets expected to fabricate the topic with the particular attributes;

Worth stream: The worth stream or worth chain is the way gone from the earliest starting point of creation to conveyance to the end client. Each progression associated with the procedure is mapped on the reason that exercises that can't be estimated can't be overseen, and those that are not recognized can't be dissected and improved. With the mapping of the worth stream it is conceivable to know and kill exercises that contain squander through waste disposal methods;

Continuous flow: From the examination and mapping of the worth flow, it is essential to make the exercises that produce worth can move through the procedure without interferences. The ideal approach to make items stream is to stream them any place conceivable by adjusting the arrangement and gear with the goal that it doesn't follow up on pausing and stock between exercises;

Pulled creation: Pulled generation means to diminish the lead time for the shopper. Executing the pulled framework implies delivering a decent or administration just when the solicitation is made by the client and not pushing the item to the customer;

Flawlessness: When the four standards are pursued plainly, that is, the association proclaims the worth precisely, maps the progression of significant worth with the goal that items stream persistently, or when clients pull these items, it is conceivable to accomplish flawlessness of procedures by killing misfortunes and waste. Consistent improvement should consistently be tried to accomplish this flawlessness. To Lean Thinking, the most significant push to flawlessness is to keep up straightforwardness among everybody engaged with the framework, so it is simpler to recognize approaches to make esteem.

The primary administration methods used to execute the Lean standards are Value Stream Mapping, Evaluation System with well-characterized measurements, 5S, Kaizen, Kanban, Standardization, Visual Management, and TPM (Total Productive Maintenance).

LSS: The Lean and Six Sigma joining

Before starting the joining among Lean and Six Sigma, it is reasonable to comprehend their likenesses and divergences, as per Antony, to upgrade crafted by the group in the truth of the association:

Similitudes:

- Center procedures in the association,
- Pertinent not just in assembling exercises,
- The board support is fundamental,
- It has a client center,
- They are comprised of multifunctional groups, and.
- The devices are integral to one another.

Divergences:

- Six Sigma requires increasingly dangerous preparing contrasted with Lean Manufacturing,
- Lean Manufacturing centers around squander decrease while Six Sigma in diminishing fluctuation,

- Six Sigma requires more significant speculation compared with Lean Manufacturing,
- Lean Manufacturing intends to streamline the progression of procedures while Six Sigma looks to expand limit,
- Lean Manufacturing doesn't present a precise system for execution, and
- Six Sigma offers explicit assignments as group strengthening.

It ought to be noticed that Six Sigma underpins Lean Manufacturing while it doesn't have an organized primary way (inborn technique) for investigating. Be that as it may, Six Sigma, like this, doesn't concentrate on improving procedure speed, decreasing lead time, and wiping out waste, which are parts of Lean Manufacturing.

Guide

The strong point for organization culture is that measurable instruments aid crafted by strategies, decreasing inconstancy and making forms increasingly steady and dependable.

As per Wiersema, there are ventures for associations to pursue, going for the association of Lean and Six Sigma. It is a guide for an incorporated usage:

Survey execution: Establish the requirement for change and evaluate how well the association is set up to roll out that improvement.

Results: Initial rundown of chances, including budgetary advantages, for ensuing prioritization and execution.

Plan the enhancements: Establish and convey the objectives of the usage of Lean Six Sigma (LSS).

Results: LSS Steering Committee, Method for determination and prioritization of tasks, Standard for estimation of monetary profits, Procedure for choice, and preparing of LSS supporters and authorities.

Empower execution: Develop, scatter, and actualize methods and strategies to build up the foundation for change.

Results: Training of backers and masters, Establishment of inner correspondence channels for the scattering of LSS, Integration of other improvement programs in power with LSS.

Execute the ventures: Execute the activities (DMAIC and Kaizen) organized.

Results: Achievement of monetary profits (approved by the controllership), Development of the "hard and delicate abilities" of the patrons and experts, Replication of activities.

Look after enhancements: Ensure the propagation of the increases accomplished and the combination of "LSS Culture," directing occasional reviews and re-invigorating the program.

Continuous improvement of LSS.

Systems for coordination

Along these lines, it is essential to remember the significance of picking the Lean Six Sigma usage methodology that best suits the qualities of the organization and its business.

Lean and Six Sigma to be coordinated into a more extensive arrangement of hierarchical improvement, the capability of which would be far more prominent than the aggregate of these two activities. In this line, a few organizations have just made their very own coordinated frameworks, called business improvement frameworks.

The technique of assembling/administrations and its vital choices with the business procedure decidedly impacts the presentation of the organization, improving efficiency and productivity, when the association has an administration model. It is conceivable to state then that for the intensity of the organization, it is significant that the decisions made in the creation of products or administrations in connection to the utilization of assets, aptitudes and improvement philosophies, for instance, ought to be guided by business techniques and chose reference models.

The most effective method to prevail with 12 fundamental lean six sigma methodologies

1. Cell fabricating

Cell fabricating is a lean, assembling approach for process improvement. Two center qualities characterize it. Gathered parts and assembling cells. In cell building, groups of elements are made in a cell of machines. A phone is a zone of a generation that is unmistakably characterized and isolated from other assembling cells – with every cell having an extreme obligation regarding the group of parts and segments. Consider cells like smaller than usual generation offices inside a bigger creation office. This is alluded to as gathering innovation.

It is an option in contrast to the conventional creation line. A creation line alludes to one continuous line of laborers that increase the value of the item from accepting the crude material to the completed item.

The significant drawback of a generation line is that an interruption in any piece of the line can end the whole procedure as every segment in the range depends on the parts that go before it.

Cell fabricating includes the re-course of action of workstations to encourage generation portrayed persistent stream and less personal time.

In the realm of assembling, all activities and machines that are expected to deliver a segment a put in closeness. Explicitly a 'U' shape. By setting the device in little assembling cells and receiving a 'U' shape, laborers invest less energy moving back and forth, between assembling parts, and additional time increasing the value of the segment.

By having hardware and workstations orchestrated in a grouping that supports rationale, you can accomplish a one-piece stream. Likewise, known as a single-piece stream and nonstop stream, the one-piece flow is the point at which your items travel through the assembling procedure at a rate controlled by the requirements of your clients.

- Tips for cell the executives
- Gathering your parts together
- Sort out your assembling cells into gatherings and sets.
- Imagine your last item as the consequence of various modules and gatherings of segments joined together

2. Takt Time

Takt Time alludes to the rate at which a completed piece is finished to satisfy client needs. It is a theoretical apparatus for recognizing if products spilling out of each station to the following in a practical way, guaranteeing that you can fulfill client needs.

In German, 'Takt' dwells in the lexical field of time and mood. In that sense, 'Takt' is the cadenced beat of your organization. Like a music conductor, Takt Time is intended to give you the way to quantify procedures to guarantee constant stream and the ideal usage of machines and systems.

The most effective method to figure Takt Time

The numerical computation for politeness time is as per the following:

The time accessible for generation ought to mirror the number of time representatives spending chipping away at the item, fewer factors, for example, gatherings, breaks, and other related exercises. On the other hand, the client request is a proportion of what number of items a client hopes to purchase.

Both of these factors ought to be predictable over a similar period, for example, at some point or seven days.

Takt Time isn't the number of worker hours put into making an item. It alludes to the whole-time length to make an item, all the way, guaranteeing that consistent stream is accomplished and client request is fulfilled.

The advantages of Takt Time

Takt Time is successfully your sell rate and is a decent estimation of how productive your work forms are. Preferably, an ideal association ought to have the limit that can undoubtedly fulfill a need without having an excess of stock in stock. Used adequately, Takt Time can Promote productivity. Your organization will have the option to gauge squander and effectively recognize which regions of creation are battling, on the plan, and generally should be balanced

- An Example of Takt Time
- Complete Time: 8 Hours X an hour = 480 Minutes
- Breaks: 50 Minutes
- Time Available: 430 Minutes
- Client Demand in 8 Hours: 100 units
- Takt Time: 430/100 = 4.3 Minutes = 258 Seconds

In this model, the client will require one unit at regular intervals. Be that as it may, you may get a kick out of the chance to create a solitary unit in minimal under 258 seconds to suit any variety in-process steps, it is essential that before you actualize takt, you guarantee that your procedures are trustworthy and can convey great quality and that your machine has an exceptionally high uptime.

3. Institutionalized Work

Institutionalized work is a basic idea. It alludes to the way toward reporting techniques, forms, materials, devices, preparing times, and the sky is the limit from there. At its center, it is tied in with guaranteeing your activities run as easily as could reasonably be expected, and your procedure improvement methodology is continually advancing and being received by your workers. Institutionalized work is imperative to arrive at your optimal Takt Time.

- Advantages of traditional work
- Best practices are pursued
- Procedure improvement never closes
- Decreases squander
- Improves scaling endeavors
- Makes anomalies progressively noticeable
- Less time spent on mystery

Tips for institutionalizing work

Guaranteeing that your workers are utilizing the prescribed procedures Is perhaps the most ideal approach to build effectiveness. If you need to advance a workplace portrayed by traditional work, you have to guarantee that your institutionalization necessities are sensible and have scope for development.

Indeed, on the off chance that you disregard the desires of representatives who will utilize these benchmarks consistently, you may wind up with a less productive workplace as the development will be smothered. Institutionalization is just wiping out elective techniques that are less proficient. At last, this implies institutionalization is increasingly fit to undertakings that are repeatable and recurrent.

- In this way, as usual, openness is of the utmost importance.
- To guarantee gauges are clung to, you have to build up to them. This involves:
- Finding a procedure or undertaking that is repeatable.
- I am setting up a perfect Takt Time for finishing this procedure.
- I am setting up the work grouping and technique that is expected to play out every component of work.
- Imparting plainly how the activity can be performed – this is ordinarily accomplished through

4. One-Piece Flow or Continuous Flow

This idea underscores lessening the cluster size to dispose of framework imperatives. A strategy by which an item or data is delivered by moving at a steady pace starting with one worth included preparing step then onto the next without any deferrals in the middle.

5. Kanban pull framework

With a Kanban pull framework, a client procedure flags a providing method to deliver an item or data when it is required.

A destroy framework alludes to JIT (Just in Time) effectiveness, where the item fulfills client need, not surpasses it. With a drawing frame, you will have a simpler time reacting to showcase powers; in any case, it is mostly about making what the client needs when they need it.

Then again, Kanban alludes to the sign utilized inside a force framework through booking joined with movement directions as straightforward visual cards and compartments.

This is in the inconsistency of 'tried and true way of thinking,' which expresses that an organization should finish items in enormous bunches. This sort of technique is known as a push framework, which is where the piece is completed before the client is prepared to get it. The significant drawback of this framework is that keeping stock costs cash, as does maintain process occupied for it.

Advantages of a Kanban pull framework

Increasingly capital – less cash will be put resources into an extra room for stock

Expanded market dynamism – Whether it is advertising powers that influence adaptability or a part of the item itself, it tends to be harming to have stock comprising of un-sellable items.

Less work in progress (WIP)

Improved generation condition – Kanban gives visual clearness and can advance target and objective discourse among colleagues

Simple checking – All colleagues will have a constant criticism of execution through a breakdown of each phase from beginning to end.

Tips for utilizing a Kanban pull framework

To effectively bring a Kanban maneuver framework into your workplace, you have to make three strides.

Guide your work process – Visualizing your work process it effectively quantifiable fragments is the center part of Kanban. Regardless of whether you utilize a physical Kanban board or an advanced variant, they commonly have three areas speaking to the condition of your item. These are; mentioned, in progress, and done.

Draw in work - When you begin to get a job, possibly pull in a new post if there is a substantial interest in it.

Oversee bottlenecks and work in progress – The motivation behind your Kanban board is to empower a smooth work process. You need to guarantee your procedures don't get stopped up by putting limits on the measure of WIP autos up at some random time.

6. Five Whys

The five why's are a proven strategy for breaking down and taking care of an issue. With the Five Why's, you can frequently get to the main driver of a problem – rather than applying a convenient solution that will eventually prompt a similar problem raising its head later on.

I am wondering why it is significant in light of the quickest method for getting to the main driver of an issue, slicing through the side effects, and getting right to the fundamental questions.

Tips on utilizing the five whys

You can use this strategy to gather an inside and out comprehension of an issue, instead of filling in the spaces yourself. This makes it incredible for investigating, yet not critical thinking.

Another factor to consider is that you can capitulate to exclusive focus and spotlight on a single reason when there could be numerous. It is continuously a smart thought of rehashing the five why test while furnishing elective responses or asking an associate to play out the five whys for examination. Therefore, you can go past five whys; the key is to stop the activity when the appropriate responses become unactionable or not any more priceless reactions are given.

Case of the five whys

- Issue - We missed a client conveyance cutoff time

- For what reason was the cutoff time missed? Since we conveyed the item one day late

- For what reason was the item conveyed late? Our client the board framework wasn't refreshed to mirror the new cluster of requests

- For what purpose wasn't the database refreshed? Since it was under the support

- For what purpose wasn't the update completed in time? There are empty positions open in the IT division, which has expanded turnaround times.

- For what reason are there negative situations in IT? A few individuals from the IT staff are on vacation simultaneously

7. Brisk Changeover/SMED

SMED is the Single Minute Exchange of Dies, which is a procedure of diminishing changeover time by ordering machine components as interior or outside, and afterward changing over the inside parts so they can be changed remotely while the machine is as yet running.

As inner changes must be performed while the machine is out of activity, the inward arrangement errands that can be changed to outer the better.

A 3-organize strategy created by Shigeo Shingo that diminishes the opportunity to change over a machine by externalizing and streamlining steps. Shorter changeover times are utilized to decrease clump sizes and produce in the nick of time. This idea helps in reducing the arrangement time to improve adaptability and responsiveness to client changes.

Advantages of SMED:

- Less vacation and improved responsiveness to clients.
- WIP and parcel size decrease.
- I have improved machine/asset use.
- By expanding the number of changeovers, we can convey less stock of crude materials, supplies, and completed merchandise.
- Become progressively productive and distinguish open doors for constant improvement.

8. Slip-up Proofing/Poka Yoke

A philosophy that keeps an administrator from making a blunder by consolidating preventive in-constructed responsiveness inside the structure of an item or creation process.

Slip-up sealing can be applied to most procedures, however, territories where it can demonstrate fundamental incorporate examples though specific process has been recognized which brings about incessant human blunder, in circumstances where the client can make a mistake when a minor error transforms into a significant error, or when anytime where a mistake will prompt substantial interruption.

Advantages of error sealing:

- Advances responsibility and procedure improvement
- Moderately low exertion and not very tedious
- Ensures that legitimate conditions exist before the real creation, and keeps deserts from occurring.
- Distinguishes and takes out reasons for interruption
- Poka-Yoke is an extraordinary method for checking blunders from developing in any way before they become more significant issues.

This procedure improvement approach is included in three stages:

- Making a flowchart of the procedure.
- Evaluating each progression

- Figuring out where there is a potential mistake discovering it at its source.
- Dispense with the wellspring of mistake or lessen its impact
- Supplant the blunder with a procedure that is mistake evidence

9. Heijunka/Leveling the Workload

The possibility that, even though client request examples might be a very factor, the entirety of our procedures should fabricate reliable amounts of work after some time (every day, hour to hour).

This technique is received by astutely arranging diverse item blend and its volumes over a time of times

10. All out Productive Maintenance (TPM)

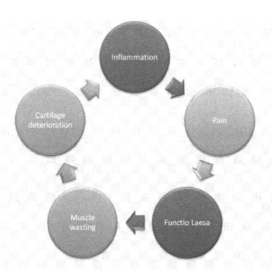

A group-based framework for improving Overall Equipment Effectiveness (OEE), which incorporates accessibility, execution, and quality. This guides in setting up a methodology for making representative possession independently for support of gear.

The objective of the TPM program is to uniquely build creation while simultaneously expanding representative confidence and employment fulfillment.

11. Five S

5S is a five-advance technique planned for making and keeping up a composed visual working environment for proceeded with process improvement and proficiency.

This is a useful framework that helps for breaking down the current hierarchical space and expelling what isn't essential.

Sifting through – This progression involves experiencing the entirety of your work apparatuses and materials to figure out what is required and what isn't. To discover the estimation of everything, ask yourself:

- What is the reason for this thing?
- What is it doing here?
- How frequently is it utilized?
- Who utilizes it?

Can't discover valuable responses to these inquiries? You presumably needn't bother with it.

Set all together – Once the pointless mess has gone, you can rework the workspace to line up with the objectives and prompt solicitations of your group.

Clear – Create an arrangement for customary upkeep and cleaning for instruments and hardware

Institutionalize – Turn one-time endeavors into propensities. Regardless of whether if it's an online agenda or verbal updates, put aside time to help encourage a domain where undertakings become schedule.

Continue – Ensure long haul supportability. Regardless of whether you're a director or a new starter, everybody should be ready for the new program. This is the reason recording techniques and guaranteeing they are anything but complicated to discover is so significant for process improvement.

12. Critical thinking/PDCA/PDSA

The PDCA/PDSA cycle is a four-eliminate graphical model for conveying change at your association. The technique is patterned, so the PDCA/PDSA cycle ought to be rehashed again and again. It is a smart thought to utilize this model toward the beginning of a procedure improvement venture, particularly for tedious forms.

Plan – Find an issue or open door as set out an arrangement for constant change. You should make a theory for what potential issues might be.

Do – This is the trying stage. Sensibly, this will be a little scale test where you can without much of a stretch measure results and increase a more prominent comprehension of your speculation

Check - Assess if the issue is fixed.

Act – If the underlying test was fruitful, rehash it on a grander scale.